From Einstein back to Newton

$$F_{Gra} = c^2 \nabla m$$

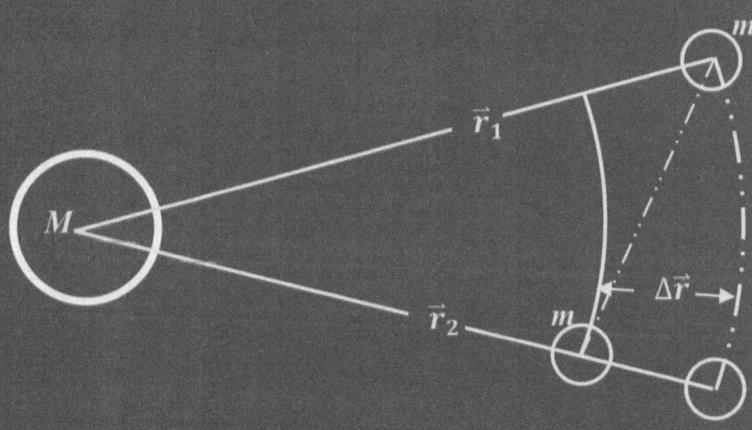

REV YUN Z QIU

From Einstein
Back to Newton

From Einstein Back to Newton

Rev. John Yun Qiu

Copyright © 2019 by Rev. John Yun Qiu.

Library of Congress Control Number:		2019919613
ISBN:	Hardcover	978-1-7960-7321-8
	Softcover	978-1-7960-7322-5
	eBook	978-1-7960-7510-6

All rights reserved. No part of this book may be reproduced or transmitted in any form or by any means, electronic or mechanical, including photocopying, recording, or by any information storage and retrieval system, without permission in writing from the copyright owner.

The views expressed in this work are solely those of the author and do not necessarily reflect the views of the publisher, and the publisher hereby disclaims any responsibility for them.

Any people depicted in stock imagery provided by Getty Images are models, and such images are being used for illustrative purposes only.
Certain stock imagery © Getty Images.

Print information available on the last page.

Rev. date: 05/01/2020

To order additional copies of this book, contact:
Xlibris
1-888-795-4274
www.Xlibris.com
Orders@Xlibris.com
804945

Preface

Newton's Law of Gravity, in spite of its deficiency, keeps its explicitness and simplicity. Do we really have to discard the Newton's Law of Gravity, instead to embrace Einstein's model of General Relativity?

The author's modified Newton's Law of Gravity successfully addressed the deficiency of Newton's Law of Gravity. The author's modified law of gravity provides equations for the orbits of the planets of the solar system that improved Newtonian orbits of the planets. The new equations depicted the motion of the perihelion of the Mercury, Venus and Earth with satisfactory results. Particularly, the Moon's period around the earth calculated via the modified Newton' Law of Gravity is more accurate than what Newton's Law of Gravity provided.

The author came with a modified law of gravity, such as $F = \frac{GMme^{\frac{\lambda}{r}}}{r^2}$, where $e^{\frac{\lambda}{r}}$ is a factor, r is the Euclidean distance between mass M and mass m, λ is a constant in respect of the corresponding central mass system. For instance, with the solar system, $\lambda = 517,000$, M is the mass of the Sun and m is the mass of a planet. In case that the earth and the moon are treated as an independent system, $\lambda = 0$ and $e^{\frac{\lambda}{r}} = 1$. Regarding the vast distance between stars in the galaxy, it does not lose generality to isolate one constellation from other constellations. The author isolated the solar system from other constellations in the Milky Way galaxy. In respect of the huge mass of the Sun and the dwarfed mass of its planets and the enormous distance between the planets, we treat the Sun as the center of the polar system and pick one planet at a time, dealing the Sun and the planet as a two-body-system, disregarding the influence from other planets. In such an idealized situation, the motion of the planet is confined in one plane because the central force (gravitational force) generates zero torque. Since the motion is a plane motion, we may apply simple polar coordinates to the system. We situate the Sun at the center of the polar coordinates, dealing it as static and the planet we picked as a subject moving around it. The position of the planet is represented by a vector $\vec{r}$.

Chapter 1

First let me introduce the modified Newton's law of gravity, such as:

$$\vec{F}_{GRA} = -\frac{GMme^{\frac{\lambda}{r}}}{r^2}\hat{r} \qquad \text{EQ [0-0]}$$

Where M is the central mass of the system, such as the sun of our solar system and $e^{\frac{\lambda}{r}}$ is the gravitational factor which is an exponential function of r, and λ is a constant depending on each specific system with our solar system.

$$\lambda = 517{,}000 = 5.17(10^5)$$

Let m be the mass of a subject of the solar system. Assuming it to be a planet and the only force working on it is the gravity of the sun, disregarding the gravitational forces from other planets.

According to Einstein's theory:

$$E = mc^2 = m_0 c^2 + W$$

Where W is the work done by the gravity of the sun, such that

$$W = \int F dr$$

$$W = \int -\frac{GMme^{\frac{\lambda}{r}}}{r^2} dr$$

$$\frac{dW}{dr} = -\frac{GMme^{\frac{\lambda}{r}}}{r^2}$$

$$dE = d(mc^2)$$

$$d(mc^2) = d(m_0 c^2 + W)$$

$$c^2 dm = dW$$

$$c^2 dm = -\frac{GMme^{\frac{\lambda}{r}}}{r^2} dr$$

$$\frac{c^2 dm}{m} = -\frac{GMe^{\frac{\lambda}{r}}}{r^2} dr$$

$$c^2 \int \frac{dm}{m} = -GM \int \frac{e^{\frac{\lambda}{r}}}{r^2} dr$$

$$c^2 (\ln m + constant) = (\frac{GM}{\lambda}) e^{\frac{\lambda}{r}}$$

Or

$$c^2 \ln m = \left(\frac{GM}{\lambda}\right) e^{\frac{\lambda}{r}} + constant$$

$$\ln m = \left(\frac{GM}{\lambda c^2}\right) e^{\frac{\lambda}{r}} + constant$$

$$\boldsymbol{m = K_0 e^{\left(\frac{GM}{\lambda c^2}\right) e^{\frac{\lambda}{r}}}} \quad \text{(Relativistic mass)} \qquad \text{EQ [0-1]}$$

Assuming

$$r = r_0, \qquad m = m_0$$

$$K_0 = m_0 e^{-\left(\frac{GM}{\lambda c^2}\right) e^{\frac{\lambda}{r_0}}}$$

Then

$$m = m_0 e^{-\left(\frac{GM}{\lambda c^2}\right) e^{\frac{\lambda}{r_0}}} e^{\left(\frac{GM}{\lambda c^2}\right) e^{\frac{\lambda}{r}}}$$

And

$$\vec{F}_{GRA} = -\frac{GMme^{\frac{\lambda}{r}}}{r^2} \hat{r}$$

$$= -GMK_0 e^{\left(\frac{GM}{\lambda c^2}\right) e^{\frac{\lambda}{r}}} e^{\frac{\lambda}{r}} \left(\frac{1}{r^2}\right) \hat{r}$$

Since we have

$$m = m_0 e^{-\left(\frac{GM}{\lambda c^2}\right) e^{\frac{\lambda}{r_0}}} e^{\left(\frac{GM}{\lambda c^2}\right) e^{\frac{\lambda}{r}}} = K_0 e^{\left(\frac{GM}{\lambda c^2}\right) e^{\frac{\lambda}{r}}}$$

$$\frac{dm}{dr} = K_0 e^{\left(\frac{GM}{\lambda c^2}\right) e^{\frac{\lambda}{r}}} \left(\frac{GM}{\lambda c^2}\right) e^{\frac{\lambda}{r}} \left(\frac{-\lambda}{r^2}\right) = -\left(\frac{GM}{c^2 r^2}\right) m e^{\frac{\lambda}{r}} = \frac{1}{c^2} F_{GRA}$$

$$\frac{c^2 dm}{dr} = F_{GRA}$$

With spherical coordinates

$$\nabla m = \frac{\partial m}{\partial r} \hat{r}, \text{ in case r is the only variable, } \nabla m = \frac{dm}{dr} \hat{r}$$

Thus

$$c^2 \nabla m = F_{GRA} \hat{r} \qquad \text{EQ [0-0*]}$$

EQ [0-1] (The relativistic mass) is justified by the period of the moon.

By EQ [2-1] as $\frac{dr}{d\theta} = 0$, $r = a$, the subject takes circular motion.

Since $\frac{d^2r}{d\theta^2} = 0$ $(\frac{dr}{d\theta})^2$ EQ (2-1) becomes $r\omega^2 = \frac{GMe^{\frac{\lambda}{r}}}{r^2}$

Now with the case of the moon and the earth

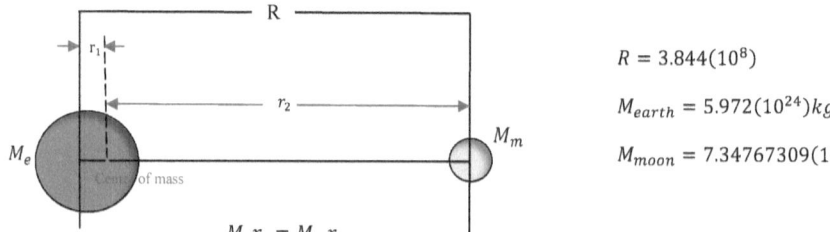

$R = 3.844(10^8)$

$M_{earth} = 5.972(10^{24}) kg$

$M_{moon} = 7.34767309(10^{22}) kg$

$M_e r_1 = M_m r_2$

$\frac{r_1}{r_2} = \frac{M_m}{M_e}$ $\frac{r_1+r_2}{r_2} = \frac{M_m + M_e}{M_e}$

$\frac{R}{r_2} = \frac{6.045476731(10^{24})}{5.972(10^{24})}$

$r_2 = \frac{3.844(10^8)(5.972)}{6.045476731}$

$r_2 = 3.79728002(10^8)$

In the case of moon and earth, we have

$\omega^2 r_2 = \frac{GM_{earth\,(relativistic)}}{R^2}$ $(e^{\frac{\lambda}{r}} = 1)$

$M_{earth\,(relativistic)} = M_{earth} e^{(\frac{GM}{c^2\lambda})e^{\frac{\lambda}{r}}}$ EQ [0-1]

Where $G = 6.67408(10^{-11})$ $M_{sun} = 1.989(10^{30}) kg$

$M_{earth} = 5.972(10^{24}) kg$ $c = 2.99792458(10^8)$

$\lambda = 517000$ $r = 1.495(10^{11})\ meters$

$e^{\frac{\lambda}{r}} = e^{\left(\frac{517000}{1.495(10^{11})}\right)} = e^{3.45819398(10^{-6})} = 1.000003458$

$$\left(\frac{GM}{c^2\lambda}\right) = 0.0028562678$$

$$\left(\frac{GM}{c^2\lambda}\right) e^{\frac{\lambda}{r}} = 0.0028562776$$

$$e^{\left(\frac{GM}{c^2\lambda}\right) e^{\frac{\lambda}{r}}} = 1.002860361$$

$$M_{earth\ (relativistic)} = M_{earth}\ (1.002860361)$$

$$\omega^2 r_2 = \frac{6.67408(10^{-11}) M_{earth}\ (1.002860361)}{[3.844(10^8)]^2}$$

$$\omega^2 = \frac{6.67408(10^{-11})(5.972)(10^{24})(1.002860361)}{[3.844(10^8)]^2 (3.79728002)(10^8)}$$

$$\omega^2 = 7.12380939(10^{-12})$$

$$\omega = 2.669046532(10^{-6})/sec$$

Earth day: 23.93 hours

Period of the moon: T = (23.93)(27.32)(3600)

$$\omega_{moon} = \frac{2\pi}{(23.93)(27.32)(3600)} = 2.669647826(10^{-6})/sec$$

With Newtonian equation

$$\omega^2 r_2 = \frac{GM_{earth}}{R^2}$$

where M_{earth} is the Newtonian mass, $M_{earth} = 5.972(10^{24})$

$$\omega^2 = \frac{GM_{earth}}{R^2 r_2} = \frac{6.67408(10^{-11}) 5.972(10^{24})}{[3.844(10^8)]^2 (3.79728002)(10^8)}$$

$$\omega^2 = 7.103490841(10^{-12})$$

$$\omega = 2.665237483(10^{-6})/sec$$

Obviously, with the modified mass of the earth, $M_{Earth\ relativistic} = M_{Earth}(1.002860361)$

$$\omega = 2.669046532(10^{-6})/sec$$

That is closer to the observed $\omega_{moon} = 2.669647826(10^{-6})/sec$ than Newton's result.

Since with Newton's unmodified mass:

$$m = 5.972(10^{24})$$

$$\omega = 2.665237483$$

It demonstrates that EQ [0-1]: $M_{Earth\ relativistic} = M_{Earth}\ e^{(\frac{GM}{c^2\lambda})e^{\frac{\lambda}{r}}}$ works.

Chapter 2

Review of vector analytic

In a planary polar coordinates, let's set the center of the coordinates at the center of the gravity of the sun. The position vector of a planet is $\vec{r}$, where $\vec{r} = r(\theta)\hat{r}$, where $r(\theta)$ is the distance from the center of the planet to the center of the sun. It is a function of the polar angle θ.

$$\vec{v} = \dot{\vec{r}} = \frac{d(r\hat{r})}{dt} = \left(\frac{dr}{dt}\right)\hat{r} + r\left(\frac{d\hat{r}}{dt}\right)$$

Where $\frac{d\hat{r}}{dt} = \dot{\hat{r}} = \vec{\omega}\hat{r} = \omega\hat{\theta}$ EQ [0-2*]

$$\frac{dr}{dt} = \dot{r} = v_r = \left(\frac{dr}{d\theta}\right)\left(\frac{d\theta}{dt}\right) = \omega\left(\frac{dr}{d\theta}\right)$$

$$\omega = \left(\frac{d\theta}{dt}\right), \quad\quad v_\theta = r\omega$$

Then

$$\dot{\hat{\theta}} = -\omega\hat{r} \quad\quad\quad\quad \text{EQ [0-3*]}$$

$$\vec{v} = \omega\left(\frac{dr}{d\theta}\right)\hat{r} + r\omega\hat{\theta} \quad\quad\quad\quad \text{EQ [0-2]}$$

$$= \omega\left[\left(\frac{dr}{d\theta}\right)\hat{r} + r\hat{\theta}\right]$$

$$|\vec{v}|^2 = \omega^2\left(\frac{dr}{d\theta}\right)^2 + r^2\omega^2$$

$$|\vec{v}| = \omega\left[\left(\frac{dr}{d\theta}\right)^2 + r^2\right]^{\frac{1}{2}}$$

$$\dot{\vec{v}} = \frac{d\vec{v}}{dt} = \frac{d}{dt}\left[\omega\left(\frac{dr}{d\theta}\right)\hat{r} + r\omega\hat{\theta}\right]$$

$$\dot{\bar{v}} = \left(\frac{d\bar{v}}{dr}\right) = \frac{d}{dt}\left[\omega\left(\frac{dr}{d\theta}\right)\hat{r} + r\omega\hat{\theta}\right]$$

$$= \left\{\left(\frac{d\omega}{dr}\right)\left(\frac{dr}{d\theta}\right) + \omega\left[\frac{d\left(\frac{d\gamma}{d\theta}\right)}{dt}\right]\right\}\hat{r} + \omega\left(\frac{dr}{d\theta}\right)\left(\frac{d\hat{r}}{dt}\right) + \frac{d(r\omega)}{dt}\hat{\theta} + r\omega\left(\frac{d\hat{\theta}}{dt}\right)$$

$$= \left\{\left(\frac{d\omega}{d\theta}\right)\left(\frac{d\theta}{dt}\right)\left(\frac{dr}{d\theta}\right) + \omega\left[\frac{d^2r}{d\theta^2}\right]\left(\frac{d\theta}{dr}\right)\right\}\hat{r} + \omega^2\left(\frac{dr}{d\theta}\right)\hat{\theta}$$
$$+ \left[\left(\frac{dr}{d\theta}\right)\left(\frac{d\theta}{dt}\right)\omega + r\left(\frac{d\omega}{d\theta}\right)\left(\frac{d\theta}{dt}\right)\right]\hat{\theta} + r\omega(-\omega)\hat{r}$$

$$= \left[\omega\left(\frac{d\omega}{d\theta}\right)\left(\frac{dr}{d\theta}\right) + \omega^2\left(\frac{d^2r}{d\theta_2}\right) - r\omega^2\right]\hat{r} + \left[2\omega^2\left(\frac{dr}{d\theta}\right) + r\omega\left(\frac{d\omega}{d\theta}\right)\right]\hat{\theta}$$

EQ [0-3]

By Newton's Second Low of Motion

$$\bar{F}_{IN} = \frac{d\bar{p}}{dt}, \quad \text{where} \quad \bar{p} = m\bar{v}$$

And

$$m = cm_o(c^2 - v^2)^{-\frac{1}{2}}$$

And Further,

$$m = \left(\frac{c}{k_1}\right)m_o e^{\left(\frac{GM}{c^2\lambda}\right)e^{\frac{\lambda}{r}}}$$

$$\bar{F}_{IN} = \frac{d\bar{p}}{dt} = \left(\frac{dm}{dr}\right)\left(\frac{dr}{dt}\right)\bar{v} + m\left(\frac{d\bar{v}}{dt}\right)$$

$$= -\left(\frac{c}{k_1}\right)m_o\left(\frac{GM}{c^2\lambda}\right)e^{\frac{\lambda}{r}}\left(\frac{\lambda}{r^2}\right)e^{\left(\frac{GM}{c^2\lambda}\right)e^{\frac{\lambda}{r}}}\left(\frac{dr}{dt}\right)\bar{v} + \left(\frac{c}{k_1}\right)m_o e^{\left(\frac{GM}{c^2\lambda}\right)e^{\frac{\lambda}{r}}}\left(\frac{d\bar{v}}{dt}\right)$$

$$= \left(\frac{c}{k_1}\right)m_o e^{\left(\frac{GM}{c^2\lambda}\right)e^{\frac{\lambda}{r}}}\left[-\left(\frac{GM}{c^2\lambda}\right)\left(\frac{\lambda}{r^2}\right)e^{\frac{\lambda}{r}}\left(\frac{dr}{dt}\right)\bar{v} + \frac{d\bar{v}}{dt}\right]$$

Where

$$\frac{dr}{dt} = \left(\frac{dr}{d\theta}\right)\left(\frac{d\theta}{dt}\right) = \omega\left(\frac{dr}{d\theta}\right)$$

And

$$\vec{v} = \dot{\vec{r}} = \omega\left(\frac{dr}{d\theta}\right)\hat{r} + \omega r\hat{\theta}$$

$$\dot{\vec{v}} = \left[\omega^2\left(\frac{d^2r}{d\theta^2}\right) + \omega\left(\frac{d\omega}{d\theta}\right)\left(\frac{dr}{d\theta}\right) - r\omega^2\right]\hat{r} + \left[2\omega^2\left(\frac{dr}{d\theta}\right) + r\omega\left(\frac{d\omega}{d\theta}\right)\right]\hat{\theta}$$

EQ [0-3]

Thus

$$\vec{F}_{IN} = e^{\left(\frac{GM}{c^2\lambda}\right)e^{\frac{\lambda}{r}}} m_0\left(\frac{c}{k_1}\right)\left[\left(\frac{-GM}{c^2r^2}\right)\omega^2\left(\frac{dr}{d\theta}\right)^2 e^{\frac{\lambda}{r}} + \omega\left(\frac{d\omega}{d\theta}\right)\left(\frac{dr}{d\theta}\right) + \omega^2\left(\frac{d^2r}{d\theta^2}\right) - r\omega^2\right]\hat{r}$$
$$+ e^{\left(\frac{GM}{c^2\lambda}\right)e^{\frac{\lambda}{r}}} m_0\left(\frac{c}{k_1}\right)\left[r\omega^2\left(\frac{-GM}{c^2r^2}\right)\left(\frac{dr}{d\theta}\right)e^{\frac{\lambda}{r}} + 2\omega^2\left(\frac{dr}{d\theta}\right) + r\omega\left(\frac{d\omega}{d\theta}\right)\right]\hat{\theta}$$

EQ [2]

(Detailed Derivation See Note 1)

With the two body system, the sun and the planet, the only force working on the planet is the gravitation force between the sun and the planet.

$$\vec{F}_{GRAV} = -\frac{GMme^{\frac{\lambda}{r}}}{r^2}\hat{r}$$

EQ [0-0]

Where

$$m = m_{10}e^{\left(\frac{GM}{c^2\lambda}\right)e^{\frac{\lambda}{r_0}}} e^{-\left(\frac{GM}{c^2\lambda}\right)e^{\frac{\lambda}{r}}}$$

And $\vec{F}_{GRA} = \vec{F}_{IN}$

Thus we obtained

$$\begin{cases} \left(\dfrac{GM}{c^2 r^2}\right)\omega^2\left(\dfrac{dr}{d\theta}\right)^2 e^{\frac{\lambda}{r}} + \omega\left(\dfrac{d\omega}{d\theta}\right)\left(\dfrac{dr}{d\theta}\right) + \omega^2\left(\dfrac{d^2 r}{d\theta^2}\right) - r\omega^2 = \dfrac{-GMe^{\frac{\lambda}{r}}}{r^2} & \text{EQ [2-1]} \\ r\omega^2\left(\dfrac{GM}{c^2 r^2}\right)\left(\dfrac{dr}{d\theta}\right)e^{\frac{\lambda}{r}} + 2\omega^2\left(\dfrac{dr}{d\theta}\right) + r\omega\left(\dfrac{d\omega}{d\theta}\right) = 0 & \text{EQ [2-2]} \end{cases}$$

By EQ [2-2], where

$$r\omega\left(\dfrac{d\omega}{d\theta}\right) = r\omega\left(\dfrac{d\omega}{dr}\right)\left(\dfrac{dr}{d\theta}\right)$$

We come with

$$-r\omega^2\left(\dfrac{GM}{c^2 r^2}\right)\left(\dfrac{dr}{d\theta}\right)e^{\frac{\lambda}{r}} + 2\omega^2\left(\dfrac{dr}{d\theta}\right) + r\omega\left(\dfrac{d\omega}{dr}\right)\left(\dfrac{dr}{d\theta}\right) = 0$$

$$\omega\left(\dfrac{dr}{d\theta}\right)\left[-r\omega\left(\dfrac{GM}{c^2 r^2}\right)e^{\frac{\lambda}{r}} + 2\omega + r\left(\dfrac{d\omega}{dr}\right)\right] = 0$$

$\omega = 0$ is a solution, since $\omega = \dfrac{d\theta}{dt}$

θ is a constant, it represents a straight line motion.

$\dfrac{dr}{d\theta} = 0$ is another solution, r is a constant.

$r = a$, it represents a circular motion with constant "a" as its radius.

Now with equation

$$-r\omega\left(\dfrac{GM}{c^2 r^2}\right)e^{\frac{\lambda}{r}} + 2\omega + r\left(\dfrac{d\omega}{dr}\right) = 0 \qquad \text{EQ [2-2*]}$$

$$r\left(\dfrac{d\omega}{dr}\right) = r\omega\left(\dfrac{GM}{c^2 r^2}\right)e^{\frac{\lambda}{r}} - 2\omega$$

$$r\left(\dfrac{d\omega}{dr}\right) = \omega\left[\left(\dfrac{GM}{c^2 r}\right)e^{\frac{\lambda}{r}} - 2\right]$$

$$\dfrac{d\omega}{\omega} = \left[\left(\dfrac{GM}{c^2 r}\right)e^{\frac{\lambda}{r}} - 2\right]\dfrac{dr}{r}$$

$$\dfrac{d\omega}{\omega} = \left(\dfrac{GM}{c^2 r^2}\right)e^{\frac{\lambda}{r}}dr - \dfrac{2dr}{r}$$

$$\int \frac{d\omega}{\omega} = \int \left(\frac{GM}{c^2 r^2}\right) e^{\frac{\lambda}{r}} dr - 2\int \frac{dr}{r}$$

$$\ln \omega = \left(\frac{-GM}{c^2 \lambda}\right) e^{\frac{\lambda}{r}} - 2 \ln r + constant$$

$$\ln \omega + \ln r^2 = \left(\frac{-GM}{c^2 \lambda}\right) e^{\frac{\lambda}{r}} + constant$$

$$\ln(\omega r^2) = \left(\frac{-GM}{c^2 \lambda}\right) e^{\frac{\lambda}{r}} + constant$$

$$\omega r^2 = K_2 e^{\left(\frac{-GM}{c^2 \lambda}\right) e^{\frac{\lambda}{r}}}$$

$$\omega r^2 e^{\left(\frac{-GM}{c^2 \lambda}\right) e^{\frac{\lambda}{r}}} = K_2 \text{, where } K_2 \text{ is a constant} \qquad \text{EQ [2-3*]}$$

By EQ [2-1], where

$$\frac{d\omega}{d\theta} = \left(\frac{d\omega}{dr}\right)\left(\frac{dr}{d\theta}\right)$$

We have

$$\left(\frac{-GM}{c^2 r^2}\right)\omega^2 \left(\frac{dr}{d\theta}\right)^2 e^{\frac{\lambda}{r}} + \omega \left(\frac{d\omega}{dr}\right)\left(\frac{dr}{d\theta}\right)^2 + \omega^2 \left(\frac{d^2 r}{d\theta^2}\right) - \omega^2 r = \frac{-GM e^{\frac{\lambda}{r}}}{r^2}$$

From EQ [2-2*] we have

$$\frac{d\omega}{dr} = \frac{\omega}{r}\left[\left(\frac{GM}{c^2 r}\right) e^{\frac{\lambda}{r}} - 2\right]$$

$$\frac{d\omega}{dr} = \omega \left[\left(\frac{GM}{c^2 r^2}\right) e^{\frac{\lambda}{r}} - \frac{2\omega}{r}\right] \text{, plugging it into EQ [2-1]}$$

We have

$$\left(\frac{-GM}{c^2 r^2}\right)\omega^2 \left(\frac{dr}{d\theta}\right)^2 e^{\frac{\lambda}{r}} + \omega \left[\omega \left(\frac{GM}{c^2 r^2}\right) e^{\frac{\lambda}{r}} - \frac{2\omega}{r}\right]\left(\frac{dr}{d\theta}\right)^2 + \omega^2 \left(\frac{d^2 r}{d\theta^2}\right) - \omega^2 r = \frac{-GM e^{\frac{\lambda}{r}}}{r^2}$$

$$\left(\frac{-GM}{c^2 r^2}\right)\omega^2 \left(\frac{dr}{d\theta}\right)^2 e^{\frac{\lambda}{r}} + \omega^2 \left(\frac{GM}{c^2 r^2}\right) e^{\frac{\lambda}{r}} \left(\frac{dr}{d\theta}\right)^2 - \frac{2\omega^2}{r}\left(\frac{dr}{d\theta}\right)^2 + \omega^2 \left(\frac{d^2 r}{d\theta^2}\right) - \omega^2 r = \frac{-GM e^{\frac{\lambda}{r}}}{r^2}$$

$$-\frac{2\omega^2}{r}\left(\frac{dr}{d\theta}\right)^2 + \omega^2 \left(\frac{d^2 r}{d\theta^2}\right) - \omega^2 r = \frac{-GM e^{\frac{\lambda}{r}}}{r^2}$$

$$-\frac{2}{r}\left(\frac{dr}{d\theta}\right)^2 + \left(\frac{d^2r}{d\theta^2}\right) - r = \frac{-GM e^{\frac{\lambda}{r}}}{r^2 \omega^2}$$ EQ [2-1*]

Since from EQ [2-3*]

$$\omega r^2 e^{\left(\frac{GM}{c^2\lambda}\right)e^{\frac{\lambda}{r}}} = K_2, \text{ we have } \omega^2 r^4 e^{\left(\frac{2GM}{c^2\lambda}\right)e^{\frac{\lambda}{r}}} = K_2^2$$

$$\omega^2 r^2 e^{\left(\frac{2GM}{c^2\lambda}\right)e^{\frac{\lambda}{r}}} = \frac{K_2^2}{r^2}$$

$$\omega^2 r^2 = \left(\frac{K_2^2}{r^2}\right) e^{\left(\frac{-2GM}{c^2\lambda}\right)e^{\frac{\lambda}{r}}}$$

Plugging it into EQ [2-1*], then we have

$$-\frac{2}{r}\left(\frac{dr}{d\theta}\right)^2 + \left(\frac{d^2r}{d\theta^2}\right) - r = \frac{-GM e^{\frac{\lambda}{r}}}{\left(\frac{K_2^2}{r^2}\right) e^{-\left(\frac{2GM}{c^2\lambda}\right)e^{\frac{\lambda}{r}}}}$$

$$-\frac{2}{r}\left(\frac{dr}{d\theta}\right)^2 + \left(\frac{d^2r}{d\theta^2}\right) - r = -GM r^2 e^{\left(\frac{2GM}{c^2\lambda}\right)e^{\frac{\lambda}{r}}} \left(\frac{1}{K_2^2}\right)$$

$$\frac{2}{r}\left(\frac{dr}{d\theta}\right)^2 - r\left(\frac{d^2r}{d\theta^2}\right) + r^2 = \left(\frac{GM}{K_2^2}\right) e^{\frac{\lambda}{r}} r^3 e^{\left(\frac{2GM}{c^2\lambda}\right)e^{\frac{\lambda}{r}}}$$

Let $\quad \frac{dr}{d\theta} = z, \quad \frac{dz}{d\theta} = \frac{d^2r}{d\theta^2} = \left(\frac{dz}{dr}\right)\left(\frac{dr}{d\theta}\right) = z\left(\frac{dz}{dr}\right)$

Then we have

$$2z^2 - rz\left(\frac{dz}{dr}\right) + r^2 = \left(\frac{GM}{K_2^2}\right) e^{\frac{\lambda}{r}} r^3 e^{\left(\frac{2GM}{c^2\lambda}\right)e^{\frac{\lambda}{r}}}$$

Let $\quad z^2 = u, \quad \frac{du}{dz} = 2z, \quad \frac{du}{dr} = 2z\left(\frac{dz}{dr}\right)$

$$2rz\left(\frac{dz}{dr}\right) = r\left(\frac{du}{dz}\right)$$

$$rz\left(\frac{dz}{dr}\right) = \frac{r}{2}\left(\frac{du}{dz}\right)$$

Then we have

$$2z^2 - \frac{r}{2}\left(\frac{du}{dr}\right) + r^2 = \left(\frac{GM}{K_2^2}\right)e^{\frac{\lambda}{r}}r^3 e^{\left(\frac{2GM}{c^2\lambda}\right)e^{\frac{\lambda}{r}}}$$

$$2u - \frac{r}{2}\left(\frac{du}{dr}\right) + r^2 = \left(\frac{GM}{K_2^2}\right)e^{\frac{\lambda}{r}}r^3 e^{\left(\frac{2GM}{c^2\lambda}\right)e^{\frac{\lambda}{r}}}$$

$$\frac{du}{dr} - \frac{4u}{r} - 2r = -\left(\frac{2GM}{K_2^2}\right)e^{\frac{\lambda}{r}}r^2 e^{\left(\frac{2GM}{c^2\lambda}\right)e^{\frac{\lambda}{r}}}$$

$$\frac{du}{dr} - \frac{4u}{r} = 2r - \left(\frac{2GM}{K_2^2}\right)e^{\frac{\lambda}{r}}r^2 e^{\left(\frac{2GM}{c^2\lambda}\right)e^{\frac{\lambda}{r}}}$$

This is a first order linear ordinary differential equation of the standard form

$$\frac{du}{dr} + P(r)u = Q(r) \text{, with solution } u = \frac{\int Q(r) e^{\int P(r)dr} dr}{e^{\int P(r)dr}}$$

Where $(r) = -\frac{4}{r}$, and

$$Q(r) = 2r - \left(\frac{2GM}{K_2^2}\right)e^{\frac{\lambda}{r}}r^2 e^{\left(\frac{2GM}{c^2\lambda}\right)e^{\frac{\lambda}{r}}}$$

$$\int P(r)dr = -4\int \frac{1}{r}dr = -4\ln r = \ln r^{-4}$$

$$e^{\int P(r)dr} = r^{-4}$$

$$\int Q(r) e^{\int P(r)dr} dr = \int r^{-4}[2r - \left(\frac{2GM}{K_2^2}\right)e^{\frac{\lambda}{r}}r^2 e^{\left(\frac{2GM}{c^2\lambda}\right)e^{\frac{\lambda}{r}}}]dr$$

$$= \int 2r^{-3} dr - \int \left(\frac{2GM}{K_2^2}\right)e^{\frac{\lambda}{r}}r e^{\left(\frac{2GM}{c^2\lambda}\right)e^{\frac{\lambda}{r}}} dr$$

$$= -r^{-2} - \left(\frac{2GM}{K_2^2}\right)\int \frac{e^{\frac{\lambda}{r}}}{r^2} e^{\left(\frac{2GM}{c^2\lambda}\right)e^{\frac{\lambda}{r}}} dr$$

$$= -r^{-2} + \left(\frac{c^2}{K_2^2}\right) e^{\left(\frac{2GM}{c^2\lambda}\right)e^{\frac{\lambda}{r}}} + \text{constant}$$

$$= -r^{-2} + \left(\frac{c^2}{K_2^2}\right) e^{\left(\frac{2GM}{c^2\lambda}\right)e^{\frac{\lambda}{r}}} + C_0$$

Then $\quad u = r^4[-r^{-2} + \left(\frac{c^2}{K_2^2}\right) e^{\left(\frac{2GM}{c^2\lambda}\right)e^{\frac{\lambda}{r}}} + C_0]$

$$u = r^2[-1 + \left(\frac{c^2}{K_2^2}\right) e^{\left(\frac{2GM}{c^2\lambda}\right)e^{\frac{\lambda}{r}}} r^2 + C_0 r^2]$$

$$\frac{du}{dr} + P(r)u = Q(r), \text{ with solution } u = \frac{\int Q(r)e^{\int P(r)dr}dr}{e^{\int P(r)dr}}$$

where $P(r) = -\dfrac{4}{r}$ and

$$Q(r) = 2r - \left(\frac{2GM}{K_2^2}\right)e^{\frac{\lambda}{r}}r^{-2}e^{\left(\frac{2GM}{c^2\lambda}\right)e^{\frac{\lambda}{r}}}$$

$$\int P(r)dr = -4\int \frac{1}{r}dr = -4\ln r = \ln r^{-4}$$

$$e^{\int P(r)dr} = r^{-4}$$

$$\int Q(r)e^{\int P(r)dr}dr = \int r^{-4}\left[2r - \left(\frac{2GM}{K_2^2}\right)e^{\frac{\lambda}{r}}r^{-2}e^{\left(\frac{2GM}{c^2\lambda}\right)e^{\frac{\lambda}{r}}}\right]dr$$

$$= \int 2r^{-3}dr - \left(\frac{2GM}{K_2^2}\right)\int e^{\frac{\lambda}{r}}r^{-2}e^{\left(\frac{2GM}{c^2\lambda}\right)e^{\frac{\lambda}{r}}}dr$$

$$= -r^{-2} - \left(\frac{2GM}{K_2^2}\right)\int \left(\frac{e^{\frac{\lambda}{r}}}{r^2}\right)e^{\left(\frac{2GM}{c^2\lambda}\right)e^{\frac{\lambda}{r}}}dr$$

$$= -r^{-2} + \left(\frac{c^2}{K_2^2}\right)e^{\left(\frac{2GM}{c^2\lambda}\right)e^{\frac{\lambda}{r}}} + \text{constant}$$

$$= -r^{-2} + \left(\frac{c^2}{K_2^2}\right)e^{\left(\frac{2GM}{c^2\lambda}\right)e^{\frac{\lambda}{r}}} + c_0$$

Then

$$u = r^4\left[-r^{-2} + \left(\frac{c^2}{K_2^2}\right)e^{\left(\frac{2GM}{c^2\lambda}\right)e^{\frac{\lambda}{r}}} + c_0\right]$$

$$u = r^2\left[-1 + \left(\frac{c^2}{K_2^2}\right)e^{\left(\frac{2GM}{c^2\lambda}\right)e^{\frac{\lambda}{r}}}r^2 + c_0 r^2\right]$$

Let $\dfrac{c}{K_2} = \zeta_1$ and $c_0 = \zeta_2$

Then

$$u = r^2\left[-1 + \zeta_1 r^2 e^{\left(\frac{2GM}{c^2\lambda}\right)e^{\frac{\lambda}{r}}} + \zeta_2 r^2\right]$$

Since $u = z^2 = \left(\dfrac{dr}{d\theta}\right)^2$

Then

$$\left(\dfrac{dr}{d\theta}\right)^2 = r^2\left[-1 + \zeta_1 r^2 e^{\left(\frac{2GM}{c^2\lambda}\right)e^{\frac{\lambda}{r}}} + \zeta_2 r^2\right]$$

$$\dfrac{dr}{d\theta} = \pm r\left[-1 + \zeta_1 r^2 e^{\left(\frac{2GM}{c^2\lambda}\right)e^{\frac{\lambda}{r}}} + \zeta_2 r^2\right]^{\frac{1}{2}} \qquad \text{EQ[3]}$$

Where ζ_1 and ζ_2 depend on boundary conditions
such as when $r = r_{Min}$, $\dfrac{dr}{d\theta} = 0$
and when $r = r_{Max}$, $\dfrac{dr}{d\theta} = 0$
Thus, ζ_1 and ζ_2 can be found by equations

$$\begin{cases} \zeta_1 r_{Min}^2 e^{\left(\frac{2GM}{c^2\lambda}\right)e^{\frac{\lambda}{r_{Min}}}} + \zeta_2 r_{Min}^2 = 1 \\ \zeta_1 r_{Max}^2 e^{\left(\frac{2GM}{c^2\lambda}\right)e^{\frac{\lambda}{r_{Max}}}} + \zeta_2 r_{Max}^2 = 1 \end{cases}$$

By McLaurin Expansion

$\zeta_1 r^2 e^{\left(\frac{2GM}{c^2\lambda}\right)e^{\frac{\lambda}{r}}}$ can be turned into a quasi polynomial form and eventually

$\left(-1 + \zeta_1 r^2 e^{\left(\frac{2GM}{c^2\lambda}\right)e^{\frac{\lambda}{r}}} + \zeta_2 r^2\right)$ is turned approximately into a quadratic form.

Finally $\dfrac{dr}{d\theta} = r\left[-1 + \zeta_1 r^2 e^{\left(\frac{2GM}{c^2\lambda}\right)e^{\frac{\lambda}{r}}} + \zeta_2 r^2\right]^{\frac{1}{2}}$

$$d\theta = \dfrac{dr}{r\left[-1 + \zeta_1 r^2 e^{\left(\frac{2GM}{c^2\lambda}\right)e^{\frac{\lambda}{r}}} + \zeta_2 r^2\right]^{\frac{1}{2}}}$$

If can be analytically solved. The solution is a quasi-conic equation (curve) such as

$$r = \frac{\varepsilon\delta}{1+\varepsilon\cos(\mu\theta+\phi_0)}, \text{ where } \varepsilon = \frac{2\mu^2 A}{\zeta_1 \lambda a_2}$$

and $\delta = \frac{1}{A}$, where A is a constant.

Detail please be referred to Chapter 3.

Note 1

Since
$$\vec{F}_{IN} = \left(\frac{c}{K_1}\right) m_0 e^{\left(\frac{2GM}{c^2\lambda}\right)} e^{\frac{\lambda}{r}} \left[-\left(\frac{GM}{c^2\lambda}\right)\left(\frac{\lambda}{r^2}\right) e^{\frac{\lambda}{r}} \left(\frac{dr}{dt}\right) \vec{v} + \frac{d\vec{v}}{dt}\right]$$

Where
$$\frac{dr}{dt} = \left(\frac{dr}{d\theta}\right)\left(\frac{d\theta}{dr}\right) = \omega\left(\frac{dr}{d\theta}\right)$$

$$\vec{v} = \omega\left(\frac{dr}{d\theta}\right)\hat{r} + \omega r\hat{\theta}$$

$$\left(\frac{dr}{d\theta}\right)\vec{v} = \omega^2\left(\frac{dr}{d\theta}\right)^2\hat{r} + \omega^2 r\left(\frac{dr}{d\theta}\right)\hat{\theta}$$

$$\frac{d\vec{v}}{dt} = \left[\omega\left(\frac{d\omega}{d\theta}\right)\left(\frac{dr}{d\theta}\right) + \omega^2\left(\frac{d^2r}{d\theta^2}\right) - r\omega^2\right]\hat{r} + \left[2\omega^2\left(\frac{dr}{d\theta}\right) + r\omega\left(\frac{d\omega}{d\theta}\right)\right]\hat{\theta}$$

Then
$$\left[-\left(\frac{GM}{c^2 r^2}\right) e^{\frac{\lambda}{r}} \left(\frac{dr}{dt}\right) \vec{v} + \frac{d\vec{v}}{dt}\right]$$

$$= -\left(\frac{GM}{c^2 r^2}\right) e^{\frac{\lambda}{r}} \omega^2 \left(\frac{dr}{d\theta}\right)^2 \hat{r} - \left(\frac{GM}{c^2 r^2}\right) e^{\frac{\lambda}{r}} \omega^2 \left(\frac{dr}{d\theta}\right)^2 \hat{\theta} + \left[\omega\left(\frac{d\omega}{d\theta}\right)\left(\frac{dr}{d\theta}\right) + \omega^2\left(\frac{d^2r}{d\theta^2}\right) - r\omega^2\right]\hat{r} + \left[2\omega^2\left(\frac{dr}{d\theta}\right) + r\omega\left(\frac{d\omega}{d\theta}\right)\right]\hat{\theta}$$

$$= \left[-\left(\frac{GM}{c^2 r^2}\right) e^{\frac{\lambda}{r}} \omega^2 \left(\frac{dr}{d\theta}\right)^2 + \omega\left(\frac{d\omega}{d\theta}\right)\left(\frac{dr}{d\theta}\right) + \omega^2\left(\frac{d^2r}{d\theta^2}\right) - r\omega^2\right]\hat{r} + \left[\left(\frac{GM}{c^2 r^2}\right) e^{\frac{\lambda}{r}} \omega^2 \left(\frac{dr}{d\theta}\right) r + 2\omega^2\left(\frac{dr}{d\theta}\right) + r\omega\left(\frac{d\omega}{d\theta}\right)\right]\hat{\theta}$$

Chapter 3

Now let's study $e^{\left(\frac{2GM}{C^2\lambda}\right)}e^{\frac{\lambda}{r}}$

By Maclaurin Expansion

$$e^x = 1 + x + \frac{x^2}{2} + \frac{x^3}{3!} + \frac{x^4}{4!} + \cdots$$

$$e^{\left(\frac{2GM}{C^2\lambda}\right)}e^{\frac{\lambda}{r}} = 1 + \left(\frac{2GM}{C^2\lambda}\right)e^{\frac{\lambda}{r}} + \left(\frac{1}{2}\right)\left(\frac{2GM}{C^2\lambda}\right)^2 e^{\frac{2\lambda}{r}} + \left(\frac{1}{6}\right)\left(\frac{2GM}{C^2\lambda}\right)^3 e^{\frac{3\lambda}{r}} + \cdots +\cdots$$

$$= 1 + \left(\frac{2GM}{C^2\lambda}\right)\left[1 + \frac{\lambda}{r} + \left(\frac{1}{2}\right)\left(\frac{\lambda}{r}\right)^2 + \left(\frac{1}{6}\right)\left(\frac{\lambda}{r}\right)^3 + \left(\frac{1}{24}\right)\left(\frac{\lambda}{r}\right)^4 + \cdots\right] +$$

$$\left(\frac{1}{2}\right)\left(\frac{2GM}{C^2\lambda}\right)^2\left[1 + \frac{2\lambda}{r} + \left(\frac{1}{2}\right)\left(\frac{2\lambda}{r}\right)^2 + \left(\frac{1}{6}\right)\left(\frac{2\lambda}{r}\right)^3 + \left(\frac{1}{24}\right)\left(\frac{2\lambda}{r}\right)^4 + \cdots\right] +$$

$$\left(\frac{1}{6}\right)\left(\frac{2GM}{C^2\lambda}\right)^3\left[1 + \frac{3\lambda}{r} + \left(\frac{1}{2}\right)\left(\frac{3\lambda}{r}\right)^2 + \left(\frac{1}{6}\right)\left(\frac{3\lambda}{r}\right)^3 + \left(\frac{1}{24}\right)\left(\frac{3\lambda}{r}\right)^4 + \cdots\right] +$$

$$\left(\frac{1}{24}\right)\left(\frac{2GM}{C^2\lambda}\right)^4\left[1 + \frac{4\lambda}{r} + \left(\frac{1}{2}\right)\left(\frac{4\lambda}{r}\right)^2 + \left(\frac{1}{6}\right)\left(\frac{4\lambda}{r}\right)^3 + \left(\frac{1}{24}\right)\left(\frac{4\lambda}{r}\right)^4 + \cdots\right]$$

$$= 1 + \left(\frac{2GM}{C^2\lambda}\right) + \left(\frac{1}{2}\right)\left(\frac{2GM}{C^2\lambda}\right)^2 + \left(\frac{1}{6}\right)\left(\frac{2GM}{C^2\lambda}\right)^3 + \left(\frac{1}{24}\right)\left(\frac{2GM}{C^2\lambda}\right)^4 + \cdots +$$

$$\frac{\lambda}{r}\left[\frac{2GM}{C^2\lambda} + \left(\frac{2GM}{C^2\lambda}\right)^2 + \left(\frac{1}{2}\right)\left(\frac{2GM}{C^2\lambda}\right)^3 + \left(\frac{1}{6}\right)\left(\frac{2GM}{C^2\lambda}\right)^4 + \cdots\right] +$$

$$\left(\frac{\lambda}{r}\right)^2\left[\left(\frac{1}{2}\right)\left(\frac{2GM}{C^2\lambda}\right) + 1\left(\frac{2GM}{C^2\lambda}\right)^2 + \left(\frac{3}{4}\right)\left(\frac{2GM}{C^2\lambda}\right)^3 + \left(\frac{1}{3}\right)\left(\frac{2GM}{C^2\lambda}\right)^4 + \cdots\right] +$$

$$\left(\frac{\lambda}{r}\right)^3\left[\left(\frac{1}{6}\right)\left(\frac{2GM}{C^2\lambda}\right) + \left(\frac{2}{3}\right)\left(\frac{2GM}{C^2\lambda}\right)^2 + \left(\frac{3}{4}\right)\left(\frac{2GM}{C^2\lambda}\right)^3 + \left(\frac{4}{4}\right)\left(\frac{2GM}{C^2\lambda}\right)^4 + \cdots\right]$$

$$\zeta_1 r^2 e^{\left(\frac{2GM}{C^2\lambda}\right)}e^{\frac{\lambda}{r}} = \zeta_1 r^2\left[1 + \left(\frac{2GM}{C^2\lambda}\right) + \left(\frac{1}{2}\right)\left(\frac{2GM}{C^2\lambda}\right)^2 + \left(\frac{1}{6}\right)\left(\frac{2GM}{C^2\lambda}\right)^3 + \right.$$

$$\left.\left(\frac{1}{24}\right)\left(\frac{2GM}{C^2\lambda}\right)^4 + \cdots\right] + \zeta_1 r\lambda\left[\left(\frac{2GM}{C^2\lambda}\right) + \left(\frac{2GM}{C^2\lambda}\right)^2 + \left(\frac{1}{2}\right)\left(\frac{2GM}{C^2\lambda}\right)^3 + \right.$$

$$\left.\left(\frac{1}{6}\right)\left(\frac{2GM}{C^2\lambda}\right)^4 + \cdots\right] + \zeta_1\lambda^2\left[\left(\frac{1}{2}\right)\left(\frac{2GM}{C^2\lambda}\right) + 1\left(\frac{2GM}{C^2\lambda}\right)^2 + \right.$$

$$\left.\left(\frac{3}{4}\right)\left(\frac{2GM}{C^2\lambda}\right)^3 + \left(\frac{1}{3}\right)\left(\frac{2GM}{C^2\lambda}\right)^4 + \cdots\right] + \zeta_1\lambda^3\left(\frac{1}{r}\right)\left[\left(\frac{1}{6}\right)\left(\frac{2GM}{C^2\lambda}\right) + \right.$$

$$\left.\left(\frac{2}{3}\right)\left(\frac{2GM}{C^2\lambda}\right)^2 + \left(\frac{3}{4}\right)\left(\frac{2GM}{C^2\lambda}\right)^3 + \left(\frac{4}{4}\right)\left(\frac{2GM}{C^2\lambda}\right)^4 + \cdots\right]\cdots$$

$$= \zeta_1\alpha_1 r^2 + \zeta_1\lambda\alpha_2 r + \zeta_1\lambda^2\alpha_3 + \zeta_1\lambda^3\left(\frac{1}{r}\right)\alpha_4 + \cdots$$

Where

$$\alpha_1 = 1 + \left(\frac{2GM}{c^2\lambda}\right) + \left(\frac{1}{2}\right)\left(\frac{2GM}{c^2\lambda}\right)^2 + \left(\frac{1}{6}\right)\left(\frac{2GM}{c^2\lambda}\right)^3 + \left(\frac{1}{24}\right)\left(\frac{2GM}{c^2\lambda}\right)^4 + \cdots$$

$$\alpha_2 = \left(\frac{2GM}{c^2\lambda}\right) + \left(\frac{2GM}{c^2\lambda}\right)^2 + \left(\frac{1}{2}\right)\left(\frac{2GM}{c^2\lambda}\right)^3 + \left(\frac{1}{6}\right)\left(\frac{2GM}{c^2\lambda}\right)^4 + \cdots$$

$$\alpha_3 = \left(\frac{1}{2}\right)\left(\frac{2GM}{c^2\lambda}\right) + 1\left(\frac{2GM}{c^2\lambda}\right)^2 + \left(\frac{3}{4}\right)\left(\frac{2GM}{c^2\lambda}\right)^3 + \left(\frac{1}{3}\right)\left(\frac{2GM}{c^2\lambda}\right)^4 + \cdots$$

$$\alpha_4 = \left(\frac{1}{6}\right)\left(\frac{2GM}{c^2\lambda}\right) + \left(\frac{2}{3}\right)\left(\frac{2GM}{c^2\lambda}\right)^2 + \left(\frac{3}{4}\right)\left(\frac{2GM}{c^2\lambda}\right)^3 + \left(\frac{4}{4}\right)\left(\frac{2GM}{c^2\lambda}\right)^4 + \cdots$$

Thus

$$-1 + \zeta_2 r^2 + \zeta_1 r^2 e^{\left(\frac{2GM}{c^2\lambda}\right)} e^{\frac{\lambda}{r}}$$

$$\sim -1 + \zeta_2 r^2 + \zeta_1 \alpha_1 r^2 + \zeta_1 \lambda \alpha_2 r + \zeta_1 \lambda^2 \alpha_3 + \zeta_1 \lambda^3 \alpha_4 \left(\frac{1}{r}\right) + \cdots$$

$$\sim -1 + \zeta_2 r^2 + \zeta_1 \alpha_1 r^2 + \zeta_1 \lambda \alpha_2 r + \zeta_1 \lambda^2 \alpha_3 + \zeta_1 \lambda^3 \alpha_4 \left(\frac{1}{r}\right) + \cdots$$

Let

$$\left(\frac{dr}{d\theta}\right)^2 = r^2 \left[-1 + \zeta_2 r^2 + \zeta_1 r^2 e^{\left(\frac{2GM}{c^2\lambda}\right)} e^{\frac{\lambda}{r}}\right]$$

Negligible

1.68103845(10⁻³³

$$\left(\frac{dr}{d\theta}\right)^2 \sim r^2 \left[-1 + (\zeta_1\alpha_1 + \zeta_2)r^2 + \zeta_1\lambda\alpha_2 r + \zeta_1\lambda^2\alpha_3 + \zeta_1\lambda^3\alpha_4\left(\frac{1}{r}\right) + \cdots\right]$$

$$\left(\frac{dr}{d\theta}\right)^2 \sim r^4 \left[-\frac{1}{r^2} + (\zeta_1\alpha_1 + \zeta_2) + \frac{\zeta_1\lambda\alpha_2}{r} + \frac{\zeta_1\lambda^2\alpha_3}{r^2} + \zeta_1\lambda^3\alpha_4\left(\frac{1}{r^3}\right)\right]$$

We let

$$\left(\frac{dr}{d\theta}\right)^2 = r^4 \left[(-1 + \zeta_1\lambda^2\alpha_3)\frac{1}{r^2} + \frac{\zeta_1\lambda\alpha_2}{r} + (\zeta_1\alpha_1 + \zeta_2)\right]$$

$$= r^4(1 - \zeta_1\lambda^2\alpha_3)\left[-\frac{1}{r^2} + \frac{\zeta_1\lambda\alpha_2}{(1-\zeta_1\lambda^2\alpha_3)r} + \frac{\zeta_1\alpha_1 + \zeta_2}{1-\zeta_1\lambda^2\alpha_3}\right]$$

$$\left(\frac{dr}{d\theta}\right)^2 = r^4(1 - \zeta_1\lambda^2\alpha_3)\left[-\left(\frac{1}{r} - \frac{\zeta_1\lambda\alpha_2}{2(1-\zeta_1\lambda^2\alpha_3)}\right)^2 + \left(\frac{\zeta_1\lambda\alpha_2}{2(1-\zeta_1\lambda^2\alpha_3)}\right)^2 + \frac{\zeta_1\alpha_1+\zeta_2}{1-\zeta_1\lambda^2\alpha_3}\right]$$

Let

$$1 - \zeta_1\lambda^2\alpha_3 = \mu^2$$

Then

$$\left(\frac{dr}{d\theta}\right)^2 = r^4\mu^2\left[-\left(\frac{1}{r} - \frac{\zeta_1\lambda\alpha_2}{2(1-\zeta_1\lambda^2\alpha_3)}\right)^2 + A^2\right]$$

Where

$$A^2 = \left(\frac{\zeta_1\lambda\alpha_2}{2(1-\zeta_1\lambda^2\alpha_3)}\right)^2 + \frac{\zeta_1\alpha_1+\zeta_2}{1-\zeta_1\lambda^2\alpha_3}$$

$$= \frac{(\zeta_1\lambda\alpha_2)^2 + 4(1-\zeta_1\lambda^2\alpha_3)(\zeta_1\alpha_1+\zeta_2)}{4(1-\zeta_1\lambda^2\alpha_3)^2}$$

Let

$$\left(\frac{1}{r} - \frac{\zeta_1\lambda\alpha_2}{2(1-\zeta_1\lambda^2\alpha_3)}\right) = x \qquad \frac{dx}{dr} = -\frac{1}{r^2} \qquad \frac{dr}{r^2} = -dx$$

Then

$$\frac{dr}{d\theta} = \mu r^2[A^2 - X^2]^{\frac{1}{2}}$$

$$\mu d\theta = \frac{dr}{r^2[A^2-X^2]^{\frac{1}{2}}} = \frac{-dx}{\sqrt{A^2-X^2}}$$

Thus

$$\int \mu d\theta = \int \frac{-dx}{\sqrt{A^2-X^2}}$$

$$\mu\theta = \cos^{-1}\left(\frac{X}{A}\right) + \phi_0$$

$$\mu\theta - \phi_0 = \cos^{-1}\left(\frac{X}{A}\right)$$

$$\cos(\mu\theta - \phi_0) = \frac{X}{A}$$

$$A\cos(\mu\theta - \phi_0) = X$$

$$A\cos(\mu\theta - \phi_0) = \frac{1}{r} - \frac{\zeta_1\lambda\alpha_2}{2\mu^2}$$

$$\frac{1}{r} = \frac{\zeta_1\lambda\alpha_2}{2\mu^2} + A\cos(\mu\theta - \phi_0)$$

$$r = \frac{1}{\frac{\zeta_1\lambda\alpha_2}{2\mu^2} + A\cos(\mu\theta - \phi_0)}$$

$$r = \frac{\frac{2(1-\zeta_1\lambda^2\alpha_3)}{\zeta_1\lambda\alpha_2}}{1 + \frac{2A(1-\zeta_1\lambda^2\alpha_3)}{\zeta_1\lambda\alpha_2}\cos(\mu\theta-\phi_0)}$$

$$r = \frac{\frac{2\mu^2}{\zeta_1 \lambda \alpha_2}}{1+(\frac{2A\mu^2}{\zeta_1 \lambda \alpha_2})\cos(\mu\theta-\phi_0)}$$

Where

$$A = \frac{\sqrt{(\zeta_1 \lambda \alpha_2)^2 + 4\mu^2(\zeta_1 \alpha_1 + \zeta_2)}}{2\mu^2}$$

And

$$\mu = (1 - \zeta_1 \lambda^2 \alpha_3)^{\frac{1}{2}}$$

By $\left(\frac{dr}{d\theta}\right)^2 = r^2\left[-1 + \zeta_1 r^2 e^{\left(\frac{2GM}{c^2\lambda}\right)e^{\frac{\lambda}{r}}} + \zeta_2 r^2\right]$

As

$$r = r_0$$

Or

$$r = r_{max} \qquad \frac{dr}{d\theta} = 0$$

Thus we have

$$\begin{cases} -1 + \zeta_1 r_0^2 e^{\left(\frac{2GM}{c^2\lambda}\right)e^{\frac{\lambda}{r_0}}} + \zeta_2 r_0^2 = 0 \\ -1 + \zeta_1 r_{max}^2 e^{\left(\frac{2GM}{c^2\lambda}\right)e^{\frac{\lambda}{r_{max}}}} + \zeta_2 r_{max}^2 = 0 \end{cases}$$

$$\begin{cases} \zeta_1 e^{\left(\frac{2GM}{c^2\lambda}\right)e^{\frac{\lambda}{r_0}}} + \zeta_2 = \frac{1}{r_0^2} & (1) \\ \zeta_1 e^{\left(\frac{2GM}{c^2\lambda}\right)e^{\frac{\lambda}{r_{max}}}} + \zeta_2 = \frac{1}{r_{max}^2} & (2) \end{cases}$$

EQ [1]-EQ [2]:

$$\zeta_1 \left[e^{\left(\frac{2GM}{c^2\lambda}\right)e^{\frac{\lambda}{r_0}}} - e^{\left(\frac{2GM}{c^2\lambda}\right)e^{\frac{\lambda}{r_{max}}}} \right] = \frac{1}{r_0^2} - \frac{1}{r_{max}^2}$$

$$e^{\frac{\lambda}{r_0}} = 1.000011239$$

$$e^{\frac{\lambda}{r_{max}}} = 1.000007405$$

$$\left(\frac{2GM}{C^2\lambda}\right)e^{\frac{\lambda}{r_0}} = 0.0057125355(1.000011239) = 0.0057125997$$

$$\left(\frac{2GM}{C^2\lambda}\right)e^{\frac{\lambda}{r_0}} = 0.0057125355(1.000007405) = 0.0057125778$$

$$e^{\left(\frac{2GM}{C^2\lambda}\right)e^{\frac{\lambda}{r_0}}} = 1.005728948$$

$$e^{\left(\frac{2GM}{C^2\lambda}\right)e^{\frac{\lambda}{r_{max}}}} = 1.005728926$$

$$\zeta_1 \times 2.23127 \times 10^{-8} = 2.67454534 \times 10^{-22}$$

$$\zeta_1 = 1.19866504 \times 10^{-14}$$

$$\zeta_2 = \frac{1}{r_0^2} - \zeta_1 e^{\left(\frac{2GM}{C^2\lambda}\right)e^{\frac{\lambda}{r_0}}}$$

$$\zeta_2 = 4.72589792 \times 10^{-22} - 1.19866504 \times\times 10^{-14} \times 1.005728948$$

$$\zeta_2 = -1.2055321 \times 10^{-14}$$

$$\lambda = 517{,}000$$

$$\frac{2GM}{C^2} = 2{,}953.380848$$

$$\frac{2GM}{C^2\lambda} = 0.0057125355$$

For Mercury

$$r_0 = 4.6(10^{10}) \qquad\qquad r_{max} = 6.982(10^{10})$$

$$\frac{1}{r_0^2} = 4.72589792(10^{-22}) \qquad\qquad \frac{1}{r_{max}^2} = 2.05135258(10^{-22})$$

$$\frac{\lambda}{r_0} = 1.123913043(10^{-5}) \qquad\qquad \frac{\lambda}{r_{max}} = 7.404755085(10^{-6})$$

$$-1 + \zeta_1 r_0^2 e^{\left(\frac{2GM}{C^2\lambda}\right)e^{\frac{\lambda}{r}}} + \zeta_2 r_0^2 = -1 + 1.19866504(10^{-14}) \times$$
$$(1.005728948)[4.6(10^{10})]^2 - 25509058.83$$

$$= -1 + 1.036252 = 0.036252$$

$$\left(\frac{2GM}{C^2\lambda}\right) = 0.0057125355$$

$$\zeta_1 r^2 e^{\left(\frac{2GM}{c^2\lambda}\right)e^{\frac{\lambda}{r}}} = \zeta_1 r^2 \left[1 + \left(\frac{2GM}{c^2\lambda}\right) + \left(\frac{1}{2}\right)\left(\frac{2GM}{c^2\lambda}\right)^2 + \left(\frac{1}{6}\right)\left(\frac{2GM}{c^2\lambda}\right)^3 + \left(\frac{1}{24}\right)\left(\frac{2GM}{c^2\lambda}\right)^4 + \right.$$
$$\left. \cdots \right] + \zeta_1 \lambda r \left[\left(\frac{2GM}{c^2\lambda}\right) + \left(\frac{2GM}{c^2\lambda}\right)^2 + \left(\frac{1}{2}\right)\left(\frac{2GM}{c^2\lambda}\right)^3 + \left(\frac{1}{6}\right)\left(\frac{2GM}{c^2\lambda}\right)^4 + \right.$$
$$\left. \cdots \right] + \zeta_1 r^2 \left[\left(\frac{1}{2}\right)\left(\frac{2GM}{c^2\lambda}\right) + 1\left(\frac{2GM}{c^2\lambda}\right)^2 + \left(\frac{3}{4}\right)\left(\frac{2GM}{c^2\lambda}\right)^3 + \right.$$
$$\left.\left(\frac{1}{3}\right)\left(\frac{2GM}{c^2\lambda}\right)^4 + \cdots\right] + \cdots$$

$$= \zeta_1 r^2 (1.005728883) + \zeta_1 \lambda r (0.0057452619) + \zeta_1 r^2 (0.002889041)$$

$$= r^2 (1.20553205)(10^{-14}) + r(3.56039525)(10^{-11}) + 9.256197814(10^{-6})$$

$\zeta_1 = 1.19866504(10^{-14})$ $\quad\quad \lambda = 517{,}000$

$\alpha_1 = 1.005728883$

$\alpha_2 = 0.0057452619$

$\alpha_3 = 0.002889041$

$\mu^2 = 1 - \zeta_1 \lambda^2 \alpha_3 = 1 - 1.19866504(10^{-14}) \times 517{,}000^2 \times 0.002889041$

$\mu^2 = 0.9999907438$

$\mu = 0.9999953719$

$\Delta\phi = \frac{2\pi}{\mu} - 2\pi = 2.90794053(10^{-5})$

In 100 years

$\quad\quad \phi = \Delta\phi(415.20973)(3600") = 43.466"$

For Venus

$$\lambda = 517{,}000$$

$$\frac{2GM}{c^2\lambda} = 0.0057125355$$

$$\lambda^2 = 2.67289(10^{-11})$$

$$\lambda^2[---] = 2.67289(10^{11}) \times 0.002791832$$

$$= -746{,}225{,}983.4$$

$$\frac{1}{r_0^2} = 8.65719123(10^{-23})$$

$$\frac{1}{r_{max}^2} = 8.42576442(10^{-23})$$

$$\frac{1}{r_0^2} - \frac{1}{r_{max}^2} = 2.314268(10^{-24})$$

$$\frac{\lambda}{r_{max}} = \frac{517000}{1.08942(10^{11})} = 4.745644471(10^{-6})$$

$$\frac{\lambda}{r_0} = \frac{517000}{1.07476(10^{11})} = 4.81037627(10^{-6})$$

$$e^{\frac{\lambda}{r_{max}}} = 1.000004746$$

$$e^{\frac{\lambda}{r_0}} = 1.00000481$$

$$-\left(\frac{2GM}{c^2\lambda}\right)e^{\frac{\lambda}{r_{max}}} = -0.0057125626$$

$$-\left(\frac{2GM}{c^2\lambda}\right)e^{\frac{\lambda}{r_0}} = 0.005712563$$

$$e^{-\left(\frac{2GM}{c^2\lambda}\right)e^{\frac{\lambda}{r_0}}} = 0.9943037227$$

$$e^{-\left(\frac{2GM}{c^2\lambda}\right)e^{\frac{\lambda}{r_{max}}}} = 0.9943037231$$

$$\mu^2 = |-1 + 4.467652065(10^{-6})| = 0.9999955323$$

$$\mu = 0.9999977661$$

$$\Delta\phi = \frac{2\pi}{\mu} - 2\pi = 1.40357406(10^{-5})$$

In 100 years

$$\phi = \Delta\phi(1.625456)(100)(3600") = 8.213212358"$$

For Earth

$$r_0 = 1.47(10^{11})$$

When

$$r_{max} = 1.52(10^{11})$$
$$\lambda = 517{,}000$$
$$\lambda^2 = 2.67289(10^{11})$$
$$\zeta_1 = 4.27780815(10^{-15})$$
$$\alpha_3 = 0.002889041$$
$$\mu^2 = 1 - \zeta_1 \lambda^2 \alpha_3$$
$$\mu^2 = 1 - 4.27780815(10^{-15}) \times 2.67289(10^{11}) \times 0.002889041$$
$$= 1 - 3.30336144(10^{-6})$$
$$= 0.9999966966$$
$$\mu = 0.9999983483$$
$$\Delta\phi = \frac{2\pi}{\mu} - 2\pi = 1.03778417(10^{-5})$$

In 100 years

$$\phi = 1.03778417(10^{-5}) \times 100 \times 3600" = 3.736023012"$$

Appendix

An alternative way to derive EQ [0-1]:

What we have from the special relativity and the modified Newton's Law of Gravity

By Einstein's special relativity

$$m = \frac{m_o}{\sqrt{1-\frac{v^2}{c^2}}} = \frac{cm_o}{(c^2-v^2)^{\frac{1}{2}}} = cm_o(c^2-v^2)^{-\frac{1}{2}}$$

EQ (0-1)

We set a polar coordinates with its center situated at the gravity center of the sun. And m is the mass of a certain planet of the solar system. $\vec{r}$ is the location vector of the planet with its origin established at the center of the sun. In EQ (0-1), v is the speed of the planet and here v is treated as a scalar. m_o is the static mass of the planet. And m is a function of it's speed.

By

$$m = cm_o(c^2-v^2)^{-\frac{1}{2}}$$

EQ (0-1)

$$\frac{dm}{dt} = \left(\frac{dm}{dv}\right)\left(\frac{dv}{dt}\right)$$

By EQ (0-1)

$$\frac{dm}{dv} = cm_o\left(-\frac{1}{2}\right)(c^2-v^2)^{-\frac{3}{2}}(-2v)$$

$$= cm_o v(c^2-v^2)^{-\frac{3}{2}}$$

By Einstein's Equation

$$E = mc^2, \text{ where } m = cm_o(c^2-v^2)^{-\frac{1}{2}}$$

$$\frac{dE}{dr} = \left(\frac{dE}{dv}\right)\left(\frac{dv}{dr}\right)$$

$$\frac{dE}{dv} = m_o c^3\left(-\frac{1}{2}\right)(c^2-v^2)^{-\frac{3}{2}}(-2v)$$

$$= m_o c^3 v (c^2 - v^2)^{-\frac{3}{2}}$$

Hence

$$\frac{dE}{dr} = m_o c^3 v (c^2 - v^2)^{-\frac{3}{2}} \left(\frac{dv}{dr}\right)$$

$$dE = m_o c^3 v (c^2 - v^2)^{-\frac{3}{2}} dv$$

Now we introduce a Modified Law of Gravity such as

$$\bar{F}_{Gravity} = -\frac{GMme^{\frac{\lambda}{r}}}{r^2} \hat{r}, \text{ where } m = cm_o (c^2 - v^2)^{-\frac{1}{2}} \quad \text{EQ (0-0)}$$

Where $e^{\frac{\lambda}{r}}$ is an exponential multiplier, where λ is a constant and r is the radius, (the distance from the sun to the planet).
As the planet moved a little bit from its previous location, in the polar coordinates, there is change of r, we call it dr. And change of the polar angle θ, we call it $d\theta$. The work done on the body (the planet) is such as

$$dw = F_{GRA} dr$$

$$dw = \frac{-GMme^{\frac{\lambda}{r}}}{r^2} dr$$

By the conservation of energy

$$dw = dE$$

We have

$$m = cm_o (c^2 - v^2)^{-\frac{1}{2}}$$

$$dw = -GMcm_o (c^2 - v^2)^{-\frac{1}{2}} e^{\frac{\lambda}{r}} \left(\frac{1}{r^2}\right) dr$$

And

$$dE = m_o c^3 v (c^2 - v^2)^{-\frac{3}{2}} dv$$

Thus

$$dw = \frac{-GMme^{\frac{\lambda}{r}}}{r^2} dr = -GMcm_o (c^2 - v^2)^{-\frac{1}{2}} e^{\frac{\lambda}{r}} \left(\frac{1}{r^2}\right) dr$$

We set up equation

$$-GMcm_o(c^2-v^2)^{\frac{1}{2}}e^{\frac{\lambda}{r}}\left(\frac{1}{r^2}\right)dr = m_o c^3 v(c^2-v^2)^{-\frac{3}{2}}dv$$

$$\frac{-GMe^{\frac{\lambda}{r}}}{r^2}dr = c^2 v(c^2-v^2)^{-\frac{3}{2}}(c^2-v^2)^{+\frac{1}{2}}dv$$

$$\frac{-GMe^{\frac{\lambda}{r}}}{r^2}dr = c^2 v(c^2-v^2)^{-1}dv$$

$$-\int\frac{GMe^{\frac{\lambda}{r}}}{r^2}dr = \int\frac{c^2 v}{c^2-v^2}dv$$

$$-\frac{GM}{c^2}\int\frac{e^{\frac{\lambda}{r}}}{r^2}dr = \int\frac{vdv}{c^2-v^2}$$

$$\left(\frac{GM}{c^2\lambda}\right)e^{\frac{\lambda}{r}} = -\frac{1}{2}\ell n(c^2-v^2) + const$$

$$e^{\left(\frac{GM}{c^2\lambda}\right)e^{\frac{\lambda}{r}}} = \frac{k_1}{\sqrt{c^2-v^2}} = \frac{k_1}{c\sqrt{1-\frac{v^2}{c^2}}}$$

$$\left(\frac{c}{k_1}\right)e^{\left(\frac{GM}{c^2\lambda}\right)e^{\frac{\lambda}{r}}} = \frac{1}{\sqrt{1-\frac{v^2}{c^2}}}$$

EQ (0-1*)

As $r = r_o$ $v = v_o$

Then

$$\frac{c}{k_1} = \frac{e^{-\left(\frac{GM}{c^2\lambda}\right)e^{\frac{\lambda}{r_o}}}}{\sqrt{1-\frac{v_o^2}{c^2}}}$$

Then

$$m_o\left(\frac{c}{k_1}\right)e^{\left(\frac{GM}{c^2\lambda}\right)e^{\frac{\lambda}{r}}} = \frac{m_o}{\sqrt{1-\frac{v^2}{c^2}}} = m$$

$$m = m_o\left(\frac{c}{k_1}\right)e^{\left(\frac{GM}{c^2\lambda}\right)e^{\frac{\lambda}{r}}}$$

EQ (1)

www.ingramcontent.com/pod-product-compliance
Lightning Source LLC
Chambersburg PA
CBHW021510210526
45463CB00002B/971